Section 1 Measureme and their errors

Topic 1 Use of SI units and their prefixes

There are seven fundamental base units that are used to measure the amount of physical quantities. These are (together with their SI unit): mass (kilogram, kg), length (metre, m), time (second, s), quantity of matter (mole, mol), temperature (kelvin, K) and electric current (ampere, A). The seventh base unit is the unit of light intensity (candela, cd), which is not included in your specification. Large and small values are dealt with by using prefixes and standard form. The table below shows the most common prefixes and their standard form equivalents.

All non-fundamental quantities (and their units), such as charge, are derived in terms of the seven fundamental quantities. In the case of speed, which is defined as distance divided by time, the unit is ms^{-1}.

Some quantities (such as force) are given a derived unit that is normally named after a famous scientist. In this case the derived unit is the newton N, where $1N = 1\,kg\,m\,s^{-2}$.

Some units have several different versions. This is normally used when the most convenient unit is not always the SI unit, usually involving very large or very small values. Two good examples of these are the kilowatt-hour, kWh, (which is a standard unit of electrical consumption equivalent to $3\,600\,000\,J$, or having a one kW electric fire heater on for one hour) and the electron-volt, eV, (which is the energy equivalent to $1.6 \times 10^{-19}\,J$, or the kinetic energy gained by one electron when it is accelerated through a potential difference of 1 V).

Prefix	T tera	G giga	M mega	k kilo	c centi	m milli	μ micro	n nano	p pico	f femto
Standard form	$\times 10^{12}$	$\times 10^{9}$	$\times 10^{6}$	$\times 10^{3}$	$\times 10^{2}$	$\times 10^{-3}$	$\times 10^{-6}$	$\times 10^{-9}$	$\times 10^{-12}$	$\times 10^{-15}$

1 **Write the following quantities using standard form and suitable prefixes. (AO1)**　　**3 marks**

a **atomic radius of a carbon atom, 0.000 000 000 07 m**

b **wavelength of a helium–neon laser, 0.000 000 6328 m**

c **mass of the P&O cruise ship *Britannia*, 141 000 000 kg**

Topic 2 Limitation of physical measurements

Random and systematic errors

A measurement error occurs when an experimental measurement is taken and the measured value is not the 'true value' of the quantity being measured. The 'true value' of measurement is the value of the measurement that would be obtained in an ideal world. There are two types of measurement error:

- **systematic errors** can happen if a measurement is consistently too large or too small. This type of error is usually caused by poor experimental technique; zero error on an instrument or poor calibration of the instrument. Systematic errors are reduced by correcting for offsets and using different methods or instruments to obtain the same value and identifying any systematic error.
- **random errors** can happen when repeating a measurement gives an unpredictable and different result. Random errors usually arise from observer (human) error; the readability of the equipment (if a value is changing constantly) or external effects on the measured item (e.g. changing temperature). Random errors are reduced by taking repeated measurements and averaging.

Quality of measurement

Several different words can be used to assess the quality of a measurement:

- **accuracy:** a measurement is accurate if it is considered to be close to the 'true value'
- **precision:** how closely a set of repeated measurements are to each other
- **repeatability:** the level of consistency of a set of repeated measurements made by the same person, in the same laboratory, using the same method
- **reproducibility:** the level of consistency of a set of repeated measurements made using the same method by different people in different laboratories
- **resolution:** the smallest observable change in the quantity being measured by a measuring instrument

Uncertainty

Uncertainty and error are not the same thing. Error refers to the difference between the measurement of a physical quantity and the 'true value' of that quantity. **Uncertainty** is a measure of the spread of the value which is likely to include the 'true value'. The **absolute** uncertainty has the same units as the measurement and represents the range of possible values of the measurement. If a repeated set of measurements is made, then the absolute uncertainty is given as half the range from the highest to the lowest value obtained.

Combining uncertainty

If measurements are being added or subtracted, you can calculate the uncertainty in the overall measurement by adding the absolute uncertainties:

$$(a \pm \Delta a) + (b \pm \Delta b) = (a + b) \pm (\Delta a + \Delta b)$$

$$(a \pm \Delta a) - (b \pm \Delta b) = (a - b) \pm (\Delta a - \Delta b)$$

If the calculated quantity is derived from the multiplication or division of the measured quantities, you can determine the combined percentage error by adding the percentage uncertainties of the individual measurements:

$$\text{fractional uncertainty} = \frac{\text{absolute uncertainty}}{\text{mean value}}$$

$$\text{percentage uncertainty} = \frac{\text{absolute uncertainty}}{\text{mean value}} \times 100\%$$

In general if $a = bc$ or $a = \dfrac{b}{c}$ then:

percentage error in a =
  percentage error in b + percentage error in c

Uncertainties in graphical data

Uncertainties in data are plotted as error bars on a graph.

General rules for plotting graphs

- The axis scales should cover at least half of the graph paper.
- Label axes with the quantity and unit in the format 'quantity/unit'.
- Plot points as either a small horizontal cross '+' or a diagonal cross '×'.
- Add error bars to represent the uncertainty. The length of each bar should be the length of the absolute uncertainty for the point.

Estimating uncertainty in gradient

To estimate the uncertainty in the gradient two additional lines of fit are drawn onto the data points. These are shown in Figure 1.

The line of best fit is drawn to the data. The point (0,0) is not usually plotted unless it is a measured data point. If there is a systematic error then the line of best fit may not pass through (0,0), allowing the identification of possible systematic errors. To calculate the uncertainty in the gradient, two further lines of fit are drawn: one representing the shallowest acceptable line of fit from the bottom of the upper error bar to the top of the lowest error bar (red line in Figure 1), and one representing the steepest acceptable line of fit from the top of the upper error bar to the bottom of the lowest error bar (blue line in Figure 1). The gradient of both of these lines is calculated, and the uncertainty is given by:

uncertainty of gradient =

$$\frac{(\text{maximum gradient} - \text{minimum gradient})}{2}$$

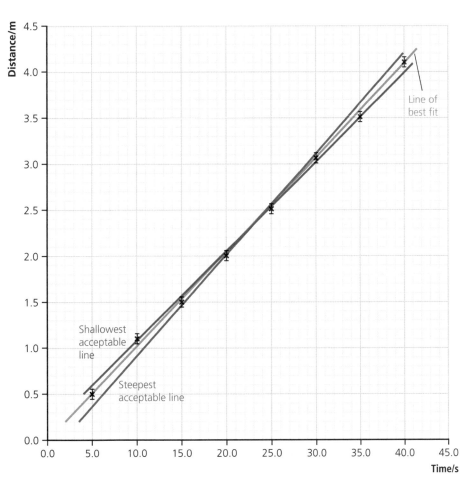

Figure 1 Estimating the uncertainty in the gradient

2 The fringe spacings measured on a screen as part of a Young's double slit experiment are shown below.

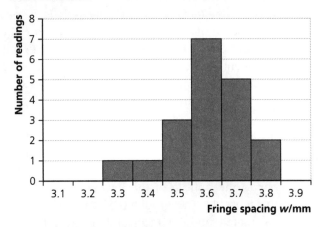

Figure 2 Fringe width distribution

Use the histogram to determine the mean average of the fringe width and use the spread of the data to determine the error on this measurement. (AO2)

Topic 3 Estimation of physical quantities

You need to be able to estimate the approximate value of quantities. You can do this easily by comparing the value that you are estimating to values that you know well. The table below gives you some common quantities that you can use to base your estimates on.

Length	Mass	Temperature
Width of a human hair ≈ 0.1 mm	Mass of an apple ≈ 100 g	Freezer temperature ≈ −20°C
Height of a door handle ≈ 1 m	Mass of a bag of sugar ≈ 1 kg	Room temperature ≈ 20°C
Length of a football pitch ≈ 100 m	Mass of a small car ≈ 1000 kg	Candle flame temperature ≈ 1000°C

Orders of magnitude

A quantity is an 'order of magnitude' bigger than another quantity, if it is about ten times bigger. Two orders of magnitude difference would be 100 times larger, or 10^2, and three orders of magnitude would be 1 000 times bigger or 10^3.

Approximated estimated values are usually given as 'orders of magnitude' to the nearest power of 10 and derived quantities can also be estimated by arithmetically combining values as orders of magnitudes.

3 **Estimate 'order of magnitude' values for the following quantities: (AO2)**   5 marks

    a   **the volume of your physics laboratory in m³**

      ....

    b   **the mass of your pencil case in kg**

      ....

    c   **your standard walking speed in m s⁻¹**

      ....

    d   **the temperature of a Bunsen burner safety flame in °C**

      ....

    e   **the breaking force of a human hair in N**

      ....

Section 2 Particles and radiation

Topic 1 Particles

Constituents of the atom

A model atom and specific charge

The model of the atom has changed considerably since it was first suggested by Democritus in about 400 BC. The model that we use today consists of a tiny central nucleus containing protons and neutrons, surrounded by a 'cloud' of orbiting electrons. A gold atom has a diameter of about 2.7×10^{-10} m whereas the gold nucleus has a diameter of about 5.4×10^{-14} m – about 5000 times smaller.

The nucleus of an atom is positively charged. This is because it contains protons that are positively charged and neutrons that are uncharged. Electrons are negatively charged, and in a neutral atom there are equal numbers of electrons and protons. The charge on the proton is $+1.6 \times 10^{-19}$ C, the neutron is neutral and the electron charge is -1.6×10^{-19} C. These numbers are difficult to visualise, so physicists give them a relative charge: +1 for protons; 0 for neutrons and –1 for electrons. The

mass of atoms and subatomic particles can be viewed in a similar way. The rest mass of a proton is 1.672×10^{-27} kg; neutrons are 1.675×10^{-27} kg and electrons are 9.11×10^{-31} kg, but these are also given in relative units; protons and neutrons are generally rounded to 1 and electrons are generally rounded to 1/1800.

The specific charge of a particle is defined as the charge per unit mass. It is an extremely useful quantity, as it forms the basis of many types of analytical techniques, such as mass spectroscopy, that are used to find the composition of substances. Any type of particle (atom, ion, nucleus or subatomic particle) can be separated by their movement in a magnetic field, which is dictated by their specific charge. Nuclei of the same element, but with different numbers of neutrons, will have different specific charges and will follow a different path inside a mass-spectrometer, allowing them to be separated and detected.

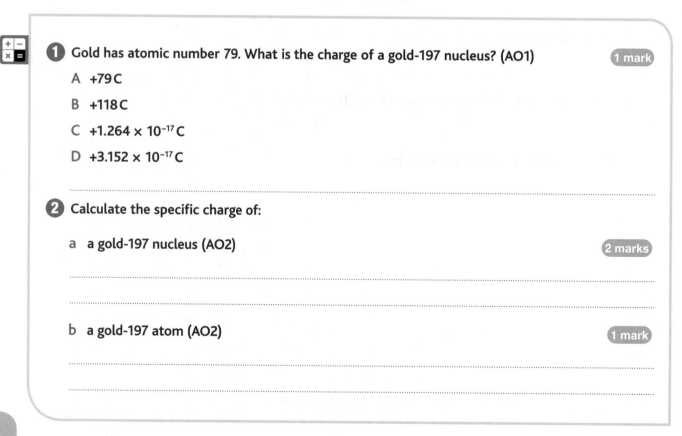

1 Gold has atomic number 79. What is the charge of a gold-197 nucleus? (AO1) `1 mark`

   A +79 C

   B +118 C

   C +1.264 × 10⁻¹⁷ C

   D +3.152 × 10⁻¹⁷ C

2 Calculate the specific charge of:

   a a gold-197 nucleus (AO2) `2 marks`

   b a gold-197 atom (AO2) `1 mark`

$^A_Z X$ notation

The proton number, Z, of a nucleus is the number of protons contained within the nucleus and the nucleon number, A, is the number of protons plus the number of neutrons in the nucleus. These two numbers uniquely identify any nucleus, and they are used together with the chemical symbol, X, of the element to describe nuclei using the $^A_Z X$ notation. Gold, for example, is $^{197}_{79}$Au.

Atoms with the same number of protons (and electrons), but different numbers of neutrons, are called isotopes and they have different $^A_Z X$ notations. Gold, for example, has 37 known different isotopes, but only one is stable ($^{197}_{79}$Au). All the others are radioactive, including $^{195}_{79}$Au which has a half-life of 186 days. $^{195}_{79}$Au has 79 protons in its nucleus but only 116 neutrons. The lightest known gold isotope is $^{169}_{79}$Au and the heaviest is $^{205}_{79}$Au.

3 Gold-198, $^{198}_{79}$Au, is a beta radiation emitter with a range in tissue of about 11 mm and a half-life of 2.7 days. It is used in some cancer treatments. Here are some numbers associated with this radioisotope:

A 198

B 79

C 119

D 0

a **What is the proton number of gold-198? (AO1)** `1 mark`

b **What is the neutron number of gold-198? (AO1)** `1 mark`

c **What is the nucleon number of gold-198? (AO1)** `1 mark`

d **What is the electron number of a gold-198 atom? (AO1)** `1 mark`

4 **State and explain the similarities and differences between the three natural isotopes of hydrogen: hydrogen, $^1_1$H; deuterium, $^2_1$H; and tritium, $^3_1$H. (AO2)** `3 marks`

Stable and unstable nuclei

Nuclei are held together by the **strong nuclear force** that acts over a very short range. It is an attractive force that operates up to a separation of about 3 fm. If nucleons are prised apart, the strong nuclear force will pull them back together, provided that they are closer than 3 fm. At separations of about 0.5 fm, the strong force becomes repulsive, stopping the nucleus from imploding. Many nuclei are stable, but some are unstable and can decay by radioactive emission. The two most common forms of radioactive decay are:

- alpha decay, where a nucleus ejects two protons and two neutrons joined together to make a helium nucleus

- beta minus decay, where a neutron decays into a proton and emits an electron and an electron antineutrino

These decays can be summarised by the following general nuclear equations:

Alpha decay: $^{A}_{Z}X \rightarrow ^{A-4}_{A-2}Y + ^{4}_{2}He$

Beta⁻ decay: $^{A}_{Z}X \rightarrow ^{A}_{z+1}Y + ^{0}_{-1}e + \bar{v}_e$

The existence of the neutrino (and the antineutrino) was hypothesised in 1930 to account for the conservation of energy during beta minus decay, but it was not experimentally observed until 1956, due to its almost negligible interaction with matter.

---

5 Write nuclear equations for the following radioactive decays:

   a polonium-210, proton number 84, via alpha decay (AO2)   `3 marks`

   b strontium-90, proton number 38, via beta minus decay (AO2)   `3 marks`

6 Use the space below to draw a labelled sketch of the variation of the strong nuclear force with the separation of a proton and a neutron. (AO1)   `5 marks`

Particles, antiparticles and photons

Every particle has a corresponding **antiparticle**. Antiparticles have the same mass and rest energy as their corresponding particle, but many of their other properties, such as their charge, have opposite values. The antiparticle of the electron, for example, is called the **positron**. The positron has the same mass–energy as an electron, but a relative charge of +1. The table below summarises some of the properties of the more common particles and their antiparticles.

Photons are the basic units of electromagnetic radiation. They behave as particles and their energy is given by the **Planck equation**:

$$E = hf \text{ or } E = \frac{hc}{\lambda}$$

where h is the Planck constant ($h = 6.6 \times 10^{-34}$ Js) and f and λ are the frequency and wavelength of the photon.

Particle–antiparticle pair	Mass/kg	Rest-energy/MeV	Relative charge/e (1e = 1.6 × 10⁻¹⁹ C)
Electron, e⁻ Positron, e⁺	9.11×10^{-31} 9.11×10^{-31}	0.511 0.511	−1 +1
Proton, p Antiproton, $\bar{p}$	1.673×10^{-27} 1.673×10^{-27}	938.3 938.3	+1 −1
Neutron, n Antineutron, $\bar{n}$	1.675×10^{-27} 1.675×10^{-27}	939.6 939.6	0 0
Electron neutrino, ν_e Electron antineutrino, $\bar{\nu}_e$	Both currently thought to be less than 3.9×10^{-36} kg	Both currently thought to be less than 2.2 eV	0 0

7 Which of the following is not an antiparticle? (AO1)   1 mark

A e⁺

B α

C $\bar{p}$

D $\bar{n}$

8 Calculate the wavelength of a 6 eV photon of electromagnetic radiation and state in which part of the electromagnetic spectrum you would find it.
($1\,eV = 1.6 \times 10^{-19}$ J) (AO2)   3 marks

Annihilation and pair creation

When a particle meets its antiparticle, they **annihilate** each other, with the creation of two gamma photons. The two gamma photons are emitted in opposite directions in order to conserve momentum. When an electron meets a positron, the total rest-energy (E_T) of the two particles is 0.511 MeV + 0.511 MeV = 1.022 MeV. This energy converts into two gamma photons, each with an energy, E, of 0.511 MeV or $(0.511 \times 10^6 \, eV \times 1.6 \times 10^{-19} \, JeV^{-1}) = 8.18 \times 10^{-14} \, J$. The wavelength of the gamma photons is given by:

$$\lambda = \frac{hc}{E} = \frac{6 \times 6 \times 10^{-34} \, Js \times 3 \times 10^8 \, ms^{-1}}{8.18 \times 10^{-14} \, J} = 2.4 \times 10^{-12} \, m$$

High-energy gamma rays can also convert into a particle–antiparticle pair when the gamma ray interacts with a large nucleus. This process is called **pair creation**. The gamma ray needs to have enough energy to create the particle and the antiparticle at rest. In order to create an electron–positron pair, where the rest energies of each particle is 0.511 MeV, the gamma ray must have at least 1.022 MeV of energy.

9 What is the combined energy released by the two photons resulting from the annihilation of a neutron–antineutron pair? (AO2) `1 mark`

A 1.022 MeV

B 1876.6 MeV

C 4.4 eV

D 1879.2 MeV

10 Muons and antimuons are heavy versions of the electron and the positron. The rest energy of a muon is 105.6 MeV. Calculate the wavelength of a gamma ray photon required to pair create a muon–antimuon pair. (AO2) `4 marks`

Particle interactions

Matter can interact in four fundamental ways, called **fundamental interactions**. Each interaction involves the interchange of exchange particles between the interacting particles giving rise to the force between them. The four fundamental interactions, their exchange particles and the nature of the interactions are shown in the table below.

The interactions between particles can be illustrated using simple diagrams called **Feynman diagrams**. Feynman diagrams show the particles before the interaction, the exchange particle and the particles after the interaction.

Particles are shown as straight lines with an arrow, indicating the order in which the particle interacts. Exchange particles are shown as wavy lines. The wavy line can have an arrow above the line indicating the order of the interaction. The vertices between the lines show the points of interaction where particles are created or annihilated.

Fundamental interaction	Exchange particle	Nature of the interaction
Electromagnetic	Virtual photon, γ	Acts between charged particles
Strong (nuclear)	Gluon, g	Acts between quarks
Weak (nuclear)	$W^{\pm}$, Z^0	Responsible for beta decay and nuclear fission
Gravity	Graviton (proposed)	Force of attraction between masses

(The gluon, graviton and the Z^0 exchange particles are not part of the examination.)

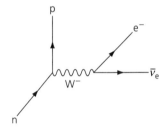

Figure 3 Feynman diagram for beta minus decay

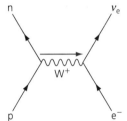

Figure 5 Feynman diagram for electron capture

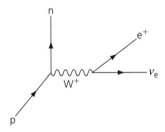

Figure 4 Feynman diagram for beta plus decay

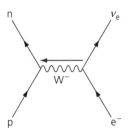

Figure 6 Feynman diagram for electron–proton collisions

11 The incomplete Feynman diagram in Figure 7 illustrates beta plus decay:

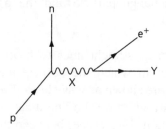

Figure 7 Feynman diagram for beta plus decay

The following table gives possible combinations for particles X and Y. Which of the combinations is correct? (AO1)

1 mark

	X	Y
A	ν_e	W^+
B	W^-	$\bar{\nu}_e$
C	$\bar{\nu}_e$	W^-
D	W^+	$\bar{\nu}_e$

12 Muons, μ^-, decay via the weak interaction into a muon neutrino, ν_μ, and a W^- exchange particle, which subsequently decays into an electron and an electron antineutrino. Draw the Feynman diagram for this particle interaction. (AO1)

5 marks

Classification of particles

Hadrons

Hadrons are particles that are subject to the **strong interaction**. There are two classes of hadrons:

- **baryons** (e.g. proton, neutron) and antibaryons (e.g. antiproton, antineutron)
- **mesons** (e.g. pion, kaon)

The proton is the only stable baryon into which all other baryons eventually decay. Baryons are assigned a quantum number called **baryon number (B)**, which distinguishes them as a baryon. All baryons have a baryon number, $B = +1$. Antibaryons have $B = -1$, and all non-baryons have $B = 0$. During all particle interactions, baryon number is conserved. This means that the total combined baryon number before the interaction must be equal to the total combined baryon number after the interaction.

The **pion** meson is the exchange particle of the strong nuclear force between nucleons (protons and neutrons). **Kaon** mesons are particles that can decay into pions. The kaon is an example of so-called strange particles that are produced through the strong interaction and decay through the weak interaction. Strange particles are assigned a **strangeness (symbol S)** quantum number to reflect the fact that strange particles are always created in pairs. Strangeness is conserved during strong interactions but it can change by 0, +1 or –1 during weak interactions.

Leptons

Leptons are particles that are subject to the **weak interaction**. The most common leptons are the electron, muon, electron-neutrino and the muon-neutrino (and their antiparticles). Muons are particles that decay into electrons. All leptons are assigned a **lepton quantum number (L)**. Leptons, such as the electron and the muon, have $L = +1$, anti-leptons, such as the positron, have $L = -1$. All non-leptons have $L = 0$.

The classification of particles by particle physicists relies on large-scale particle accelerators and detectors and on the collaborative efforts of large teams of scientists and engineers.

---

13 What is the baryon number of an alpha particle, α? (AO1) `1 mark`

A 4

B 2

C 0

D –2

---

14 Kaons are particles that are created during strong interactions, where strangeness is conserved, but decay via the weak interaction, where strangeness can remain unchanged or change by ±1. Use the strangeness quantum number data to show that the following equations could be examples of kaon production and decay. (AO2) `3 marks`

Particle	Proton	Kaon⁺	Kaon⁻	Muon	Muon antineutrino
Strangeness (S)	0	+1	−1	0	0

Production: $p + p \rightarrow p + p + K^+ + K^-$

Decay: $K^- \rightarrow \mu^- + \bar{\nu}_\mu$

Quarks and antiquarks

Hadrons are particles composed of **quarks**. Quarks are particles that appear to be fundamental building blocks of the universe (along with leptons and exchange particles). Baryons are made up of three quarks (antibaryons are made up of three antiquarks) and mesons are made up of a quark–antiquark pair. There are six quarks: up (u), down (d), strange (s), charm (c), bottom (b) and top (t), although you only need knowledge of the first three (up, down and strange) for the examinations. Each quark has a corresponding antiquark. The up, down and strange quarks have the properties shown in table below. (The antiquarks have opposite properties.)

Quark	Relative charge	Baryon number	Strangeness
Up (u)	+2/3	+1/3	0
Down (d)	−1/3	+1/3	0
Strange (s)	−1/3	+1/3	−1

There are many different combinations of three quarks, hence there are many different baryons. Two baryons (and their antibaryons), the proton and the neutron (and the antiproton and the antineutron) are much more common than any others. The quark combinations of these four particles is given below:

Baryon	Proton	Neutron	Antiproton	Antineutron
Quark composition	uud	udd	ūūd̄	d̄d̄d̄

Mesons are unstable particles that are composed of quark–antiquark pairs. The longest-living mesons only exist for a few tens of nanoseconds, and they tend to decay into electrons, neutrinos and photons. Mesons are high-energy particles that are only seen in high-energy particle interactions such as the interaction of cosmic rays with the upper atmosphere. As with baryons there are many different mesons, but the kaon and the pion are the most common. There are three varieties of kaons and pions, and their quark structures are shown below:

Kaon	K+	K⁰	K⁻
Quark structure	us̄	ds̄	su̅

Pion	π⁺	π⁰	π⁻
Quark structure	ud̄	uu̅ or dd̄	du̅

During beta minus radioactive decay, a neutron decays into a proton, an electron and an electron antineutrino. In fundamental terms involving quark structure, a down (d) quark decays into an up (u) quark and a W⁻ exchange particle. The W⁻ exchange particle then decays into an electron and an electron-antineutrino. The Feynman diagram for this decay is shown in Figure 8.

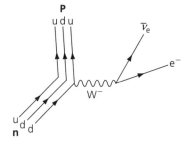

Figure 8 Feynman diagram for beta minus decay in terms of quarks

---

⑮ **The lambda⁰ baryon has the quark structure uds. Which of the following gives the combination of charge, baryon number and strangeness of the lambda⁰ baryon? (AO1)**

	Q	B	S
A	+1	−1	−1
B	−1	−1	+1
C	0	+1	−1
D	0	+1	+1

16 Figure 9 shows an unlabelled Feynman diagram for beta plus decay. During this decay a proton decays into a neutron and a W⁺ exchange particle, which subsequently decays into a positron and an electron neutrino. Label the diagram correctly.   5 marks

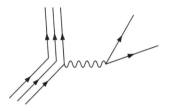

Figure 9 Unlabelled Feynman diagram for beta plus decay in terms of quarks

Applications of conservation laws

During all particle interactions energy and momentum are always conserved. **Charge (Q), baryon number (B) and lepton number (L) are also always conserved in particle interactions.** Strangeness (S) is conserved in strong interactions but can change by +1, –1, or not at all during weak interactions. Analysis of the Q, B, L and S quantum numbers can be used to determine if a particle interaction is possible or not. As examples, consider the following interactions. Example 1 shows the interactions for a neutron. Example 2 shows interactions for a proton.

Example 1: $^{1}_{0}n \rightarrow \, ^{1}_{1}p + \, ^{0}_{-1}e^{-} + \bar{v}_{e}$

Conservation quantity	Before interaction		After interaction				Quantity conserved?
	n	Total	p	e^{-}	$\bar{v}_{e}$	Total	
Q	0	0	+1	−1	0	0	✓
B	+1	+1	+1	0	0	+1	✓
L	0	0	0	+1	−1	0	✓
S	0	0	0	0	0	0	✓

In this example, all four quantities are conserved and therefore the interaction is possible.

Example 2: $^{1}_{1}p \rightarrow \, ^{1}_{0}n + \, ^{0}_{1}e^{+} + \bar{v}_{e}$

Conservation quantity	Before interaction		After interaction				Quantity conserved?
	p	Total	n	e^{+}	$\bar{v}_{e}$	Total	
Q	+1	+1	0	+1	0	+1	✓
B	+1	+1	+1	0	0	+1	✓
L	0	0	0	−1	−1	−2	✗
S	0	0	0	0	0	0	✓

In this example, lepton number is not conserved and therefore the interaction is not possible.

17 **Use the following quantum number conservation tables to determine which of the following interactions is possible. (Σ^{0} has the quark structure uds.) (AO3)**

a $p + e^{+} \rightarrow e^{-} + \Sigma^{0} + K^{+}$ `5 marks`

Conservation quantity	Before interaction			After interaction				Quantity conserved?
	p	e^{+}	Total	e^{-}	Σ^{0}	K^{+}	Total	
Q								
B								
L								
S								

Interaction possible/not possible

b $p + \bar{v}_e \rightarrow e^+ + n$

5 marks

Conservation quantity	Before interaction			After interaction			Quantity conserved?
	p	$\bar{v}_e$	Total	e^+	n	Total	
Q							
B							
L							
S							

Interaction possible/not possible

Exam-style questions

1 The table below shows some basic information about three particles:

Sub-atomic particle	Quark structure	Baryon *or* meson *or* lepton	Relative charge	Baryon number	Lepton number	Strangeness
		Meson		0		
	uud					
Positron	n/a		+1			0

a   Complete the table. **3 marks**

b   Each of the subatomic particles shown in the table has a corresponding antiparticle. State one example of a baryon and its antibaryon, NOT shown in the table, and state their quark structures. **4 marks**

....

....

....

c   The electron and the positron are an example of a lepton particle–antiparticle pair. State one property of a positron that is the same as an electron, and one property that is different. **2 marks**

....

....

....

d   A uud particle meets its antiparticle and annihilates. Write a nuclear equation for this decay and calculate the wavelength of the photons emitted. **3 marks**

....

....

e   The positron is formed during the beta plus decay of carbon-11 ($Z = 6$) to boron-11.

   i   Write a nuclear equation for this decay. **3 marks**

ii Draw a Feynman diagram for this decay.   5 marks

iii Calculate the specific charge of the boron-11 nucleus.   4 marks

....

....

....

....

f   A student proposes that the K$^-$ meson decays via the weak interaction to a muon (μ^-) and a muon antineutrino, $\bar{\nu}_\mu$. In the space below, construct a quantum number conservation table and use it to determine if this decay is possible.   5 marks

2 a Leptons and hadrons are groups of particles that make up the majority of particles in the observable universe. Describe how particles are assigned into either the lepton group or the hadron group. Your description should include the following:

- how the type of interaction is used to classify particles

- one example of each group of particle

- details of any similarities between leptons and hadrons

- details of how one group can be further sub-divided into two sub-groups

The quality of your written communication will be assessed in your answer.

6 marks

b **Leptons and hadrons consist of particles and their antiparticles.**
Complete the following table of particles and their antiparticles and
assign the particle-antiparticles to their correct group. `3 marks`

Particle	Antiparticle	Lepton or hadron
Electron-neutrino		
	Positron	
Neutron		

c **There is one other group of particles NOT mentioned in this question, that is responsible**
for the interactions between particles.

i **State the name of this group.** `1 mark`

....

ii **Complete the table identifying two of these particles and the interaction that they**
are responsible for. `2 marks`

Particle	Particle interaction involved

Topic 2 Electromagnetic radiation and quantum phenomena

Photoelectric effect

When photons of electromagnetic radiation are incident on a metal surface, the energy of the photon can be converted to kinetic energy of free electrons at the surface of the metal. If the kinetic energy of an electron is high enough, the electron can leave the surface – this is called the **photoelectric effect**. Albert Einstein studied this process in the early part of the twentieth century, and realised that the incident photons had to have a **particular** threshold frequency. Below this frequency, **photoelectrons** cannot be emitted because the electrons are held to the surface with a characteristic energy called the **work function** (ϕ). The incident photons have to have an energy greater than the work function before photoelectrons are emitted. The energy of the incident photons is given by the Planck equation, $E = hf$, and the maximum kinetic energy of any emitted photoelectrons, E_{kmax} is given by Einstein's photoelectric effect equation:

$$hf = \phi + E_{kmax}$$

The maximum kinetic energy of the photoelectrons can be determined by applying a reverse electrical potential onto the surface of the metal, called the **stopping potential** (V_s). The electrical potential energy, equal to the maximum kinetic energy of the photoelectrons, is given by $E_{kmax} = eV_s$ where e is the charge on the electron. The photoelectric effect equation can therefore be written as:

$$hf = \phi + eV_s$$

Measuring the stopping potential for different frequencies of incident electromagnetic radiation, allows experimental measurement of the work function and the Planck constant, h.

---

1 The work function of a metal surface is 2.0 eV. What is the maximum kinetic energy of photoelectrons emitted by the metal if photons with an energy of 1.9 eV are incident on the surface? (AO2)   1 mark

A 0.1 eV

B 3.9 eV

C 0 eV

D 1.9 eV

---

2 The work function of sodium is 2.28 eV. Photoelectrons are stopped from leaving the surface by a stopping voltage of 3.02 V. Calculate the wavelength of the photons responsible for this photoelectric emission. (AO2)   3 marks

Collisions of electrons with atoms

Electrons are normally held securely to atoms, but if enough energy is given to them, via a collision with a free electron, for example, then the electron can leave the atom. This is called **ionisation** as it leaves the atom with an overall positive charge, forming an ion. The energy required to do this is called the ionisation energy. If an electron around an atom absorbs an amount of energy less than the ionisation energy, then the electron can become **excited**. This means that the electron absorbs the extra energy, moving it to a higher energy level within the atom. Excited electrons normally lose their excess energy relatively quickly and return to their lowest energy level, emitting the excess energy as electromagnetic photons.

Inside fluorescent tubes, an electrical current is passed through a low-pressure mercury vapour. Collisions between the electrons of the electrical current and the mercury atoms cause electrons to be excited. These electrons subsequently lose their energy by emitting ultraviolet photons. The UV photons hit a special paint coating on the inside of the tube, which fluoresces and produces visible light photons.

The energy of electrons can be measured using **joules (J)** or **electron-volts (eV)**. One electron-volt converts to an energy of 1.6×10^{-19} J and 1 J is equivalent to 6.25×10^{18} eV. The ionisation energy of hydrogen is 13.6 eV or 21.76×10^{-19} J.

③ **The ionisation energy of helium is 24.6 eV. The wavelength of the electromagnetic radiation needed to ionise helium is: (AO2)** `1 mark`

A 50 nm

B 80 nm

C 20 μm

D 20 nm

④ **The electron in an atom of hydrogen is ionised by a photon of light with a frequency of 3.6×10^{15} Hz. Calculate the kinetic energy of the ionised electron in joules, J. (AO2)** `4 marks`

Energy levels and photon emission

Electrons exist only in discrete energy levels around the nuclei of atoms. Electrons can move up to higher energy levels either by absorbing electromagnetic radiation photons or by collisions with other atoms or free electrons. Electrons can drop down to lower energy levels by emitting electromagnetic radiation photons. The energy of a photon emitted or absorbed by an electron level transition is given by the equation:

$$hf = E_1 - E_2$$

where $E = hf$ and E_1 and E_2 are the potential energies of the two energy levels involved.

The line spectra emitted by atoms is direct evidence for these energy levels. For each element, only photons of specific frequencies (and hence energies) are emitted, corresponding to specific energy level transitions. The most common atom in the universe, hydrogen, emits three different groups of photons (an ultraviolet series, a visible series and an infrared series) corresponding to energy level transitions back to the first, second and third energy levels respectively.

5 The energy levels of hydrogen are shown in Figure 10.

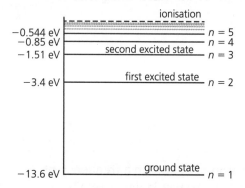

Figure 10 The energy levels of hydrogen

An electron in the $n = 1$ energy level can be excited to a higher energy level by the absorption of a photon of suitable energy. One such photon has an energy of 12.9 eV. The most probable energy level that the electron will move to would be: (AO2) `1 mark`

A $n = 2$

B $n = 3$

C $n = 4$

D $n = 5$

6 Electrons in a gas of hydrogen atoms are excited to the $n = 4$ (−0.85 eV) energy level. Calculate the wavelengths of the highest and lowest energy photons emitted by the gas. (AO2) `4 marks`

Wave–particle duality

A beam of electrons fired at a thin film of graphite produces a diffraction pattern on a fluorescent screen on the other side of the graphite target. The diffraction pattern is similar to the diffraction pattern produced by light when it passes through a diffraction grating. Diffraction is a property of waves, so in this instance the electrons are behaving as waves. However, other experiments (such as the detection of beta-particles (electrons) by a Geiger counter) and the photoelectric effect clearly show electromagnetic waves behaving as particles. Electrons, like photons and other particles, can behave as particles or waves on the quantum scale. This **wave–particle duality** was first studied in detail by the French physicist Louis de Broglie in 1924. He proposed that particles on the quantum scale have a wavelength (λ) given by the equation:

$$\lambda = \frac{h}{mv}$$

where mv is the momentum of the particle, and h is the Planck constant. De Broglie presented his work as part of his PhD dissertation and his examiners were so unsure of his ideas that they referred his thesis to Albert Einstein for peer review and validation. Einstein had begun working on quantum theory about twenty years earlier than de Broglie and was impressed by de Broglie's thesis. Einstein duly recommended that de Broglie pass his PhD, and de Broglie's work at a stroke improved our understanding of the nature and behaviour of matter.

The amount of diffraction produced by an electron beam–graphite target system depends on the relative size of the **de Broglie wavelength** of the electrons (λ) and the interatomic spacing of the graphite atoms (d). When $\lambda \approx d$, there is the most diffraction and the pattern is spread out. As the wavelength increases (by decreasing the momentum and hence velocity of the electrons), the amount of diffraction decreases and the pattern gets narrower.

7 Which of the following will increase the wavelength of an electron undergoing electron diffraction: (AO1) `1 mark`

A increasing the separation of the atoms in the crystal

B decreasing the momentum of the electrons

C increasing the speed of the electrons

D increasing the momentum of the electrons

8 The interatomic spacing of atoms in crystals of sodium chloride is 564 pm. Calculate the speed of electrons, of mass 9.11×10^{-31} kg, required to produce maximum diffraction from sodium chloride crystals. (You may neglect relativistic effects.) (AO2) `3 marks`

Exam-style questions

1 When monochromatic light is shone on to a cadmium surface inside a photocell, photoelectrons with a maximum kinetic energy of 3.51×10^{-20} J are released. The work function of cadmium is 4.07 eV.

a   i   State what is meant by the work function of the metal surface.   **2 marks**

    ii   Explain why the emitted electrons will have a range of kinetic energies up to a maximum value.   **4 marks**

    iii   Calculate the wavelength of the light. Give your answer to an appropriate number of significant figures.   **4 marks**

b In order to explain the observations of the photoelectric effect, the wave model of electromagnetic radiation was replaced by the photon model. Explain what must happen in order for an existing scientific theory to be modified or replaced with a new theory.

2 marks

....

....

....

c The electrons emitted with maximum kinetic energy have a wave–particle duality and can behave as waves with a characteristic wavelength. Calculate the wavelength of these electrons.

3 marks

....

....

....

....

....

....

....

d i The ionisation energy of neutral free cadmium atoms is −8.99 eV. Explain what is meant by ionisation energy.

2 marks

....

....

....

ii The energy required to remove a second electron from an ion of cadmium is −16.91 eV. Electrons in this energy level can be excited from this energy level to the −8.99 eV energy level. Explain why excitation is different to ionisation.

1 mark

....

....

iii Photons of electromagnetic radiation can be used to excite electrons in the lower energy level into the higher energy level. Calculate the wavelength of the electromagnetic radiation required to do this.

3 marks

....

....

....

....

....

....

2 The spectroscope was invented by Joseph Fraunhofer in 1814. He used his apparatus to observe and measure the line spectra produced by light from flames. At the time these spectral lines could not be explained by theory. Our modern model of atoms explains how photons of characteristic frequency are emitted when atoms are bombarded with electrons. The spectrum of the light emitted contains emission lines, each of a definite wavelength. Explain how:

- the bombarding electrons cause the atoms of the gas to emit photons

- the existence of a spectrum consisting of emission lines of a definite wavelength is experimental evidence for the model that atoms have discrete energy levels

The quality of your written communication will be assessed in this question.

6 marks

Section 3 Waves

Topic 1 Progressive and stationary waves

Progressive waves

Wave terminology

Progressive waves are waves that transmit energy from one place to another. In the case of mechanical waves such as water waves, seismic waves or the waves on a slinky spring, energy is transferred by the oscillation of particles of a medium. In the case of electromagnetic waves, the passage of a combined oscillating electric and magnetic field transfers the energy.

The main terminology used to describe waves is illustrated by the diagram below, showing two transverse waves, one drawn with respect to time and the other in terms of distance.

amplitude (A) — the maximum displacement of the wave, measured from the midpoint of the oscillation (called the wave axis)

frequency (f) — the number of oscillations of the wave per second, measured in hertz (Hz)

wavelength (λ) — the distance between consecutive points on the wave that are in phase, such as two crests or two troughs, measured in metres (m)

wave speed (c) — the distance travelled by the wave per second, measured in $\mathrm{m\,s^{-1}}$

period (T) — the time taken for one complete wave oscillation, measured in seconds (s)

Wave speed, frequency and wavelength are related to each other through the basic wave equation:

$$c = f \times \lambda$$

Frequency and period are related through:

$$f = \frac{1}{T}$$

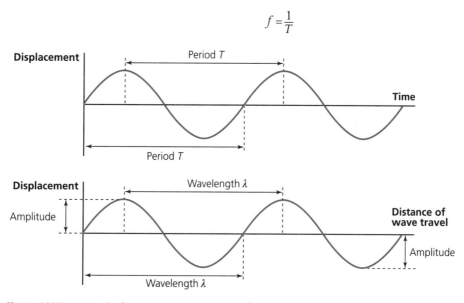

Figure 11 Wave terminology

Phase and phase difference

The **phase** of a wave describes the fraction of the wave cycle that has elapsed since the origin of the wave. Phase is usually measured as an angle, either in degrees

(where one complete cycle is 360°), or in radians (where one complete cycle is 2π radians), or by the fraction of the cycle. For example, a wave that is a quarter of the time through its cycle, has a phase of 90°, $\pi/2$ radians or $T/4$ cycle.

31

Two points on a wave, or two different waves, can be compared by their **phase difference**. This is the difference in the phase angle or fraction of the cycle between the two points. Two points with the same phase are said to be in-phase and two points that are exactly half a cycle or 180° or π radians apart are said to be in anti-phase.

1 Water waves in a ripple tank are produced by a dipper with a period of 1.5 s and have a wavelength of 5 mm. The dipper's frequency is halved. What is the new wavelength of the waves? (AO2/AO1)

1 mark

A 5 mm

B 10 mm

C 2.5 mm

D 25 mm

2 The diagram below shows three buoys on the surface of the water inside a harbour as a wave from the wake of a boat passes through the array of buoys.

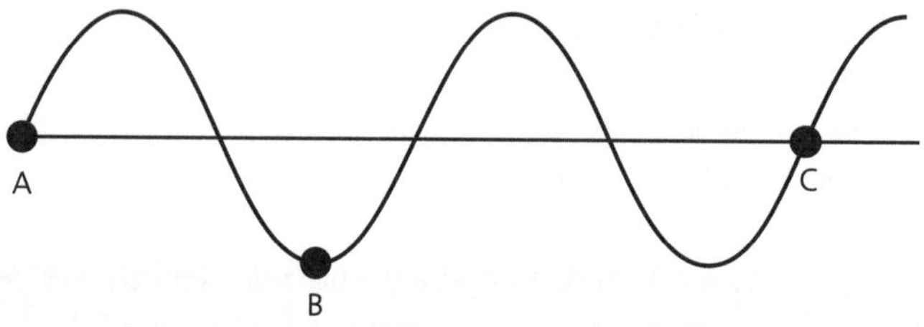

Figure 12 Buoys on water

a Stating your answer in radians, what is the phase difference between the following buoys? (AO2)

3 marks

A and B

A and C

B and C

b Which two buoys are in phase? (AO1)

1 mark

c The wave takes 5.2 s travelling at a speed of 0.6 m s⁻¹ to travel from buoy A to buoy C. Calculate the wavelength of the wave. (AO2)

2 marks

Longitudinal and transverse waves

Figure 13 shows the difference between **longitudinal** and **transverse** waves on a slinky spring.

With longitudinal waves, such as sound, p-seismic waves and longitudinal slinky spring waves, the direction of displacement of the particles is in the same line as the direction of energy propagation. For transverse waves, such as all the electromagnetic waves, s-seismic waves, waves on a string and transverse slinky waves, the direction of displacement of the particles or the fields is at right angles to the direction of energy propagation. All electromagnetic waves travel at the same speed in a vacuum, the speed of light, $c = 3.0 \times 10^8\,\mathrm{m\,s^{-1}}$.

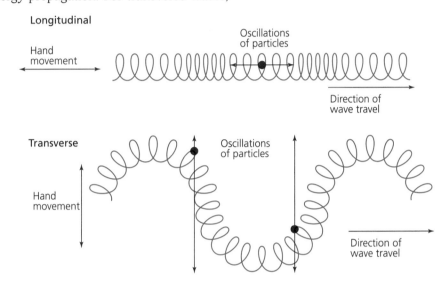

Figure 13 Longitudinal and transverse waves on a slinky spring

Polarisation

Transverse waves can be **polarised** by a suitable polariser — a property only of transverse waves. Non-polarised electromagnetic waves have their electric and magnetic fields propagating in all directions. If the wave is passed through a polariser, the material of the polariser absorbs the components of the fields in the direction of the polariser. Components of the field at right angles to the polariser travel through unaffected, leaving only one plane of polarisation, as shown in Figure 14.

Polarising sunglasses have lenses with polarising (Polaroid) filters to reduce the glare of light reflecting from horizontal surfaces such as water and snow. This reduces eyestrain for skiers and makes vision safer for drivers. Radio and television aerials also work by utilising the polarisation effect — transmitters generate plane-polarised electromagnetic waves, which are picked up most effectively by a receiver with the same plane of polarisation.

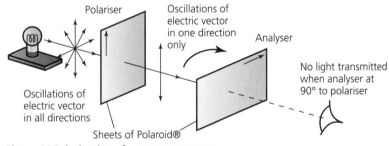

Figure 14 Polarisation of a transverse wave

3 Which of the following is a longitudinal wave? (AO1)

   A   ultrasound

   B   visible light

   C   s-seismic waves

   D   infrared

4 a State one difference and one similarity between a transverse and longitudinal wave. (AO1)

   b The diagram below shows two identical polarising filters, A and B, and an unpolarised light bulb source. The arrows indicate the plane in which the electric field of the light oscillates.

     i Polarised light reaches the observer. Draw the direction of the transmission axis on filter B in Figure 15.

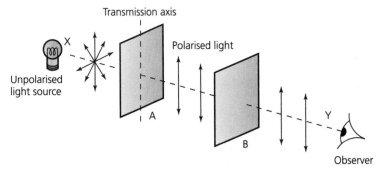

Figure 15

     ii The polarising filter B is rotated anticlockwise through 360° about line XY from the position shown in Figure 16. On the graph axes below, sketch how the light intensity reaching the observer varies with angle. (AO2)

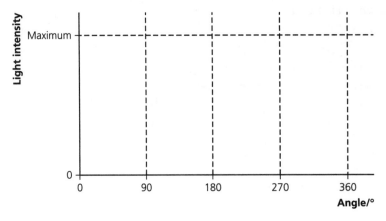

Figure 16

   c Explain why polarising filters are used in sunglasses. (AO1)

Principle of superposition of waves and formation of stationary waves

Stationary waves are formed when two waves of the same frequency travel in opposite directions and interfere with each other, as shown in Figure 17.

The resulting stationary wave exists as a series of nodes and antinodes. Nodes are places of zero amplitude and antinodes are places of maximum amplitude. A vibrating string can produce a family of different stationary waves with frequencies given by the equation:

$$f = \frac{1}{2l}\sqrt{\frac{T}{\mu}}$$

where l is the length of the string, T is the tension in the string (measured in newtons, N) and μ is the mass per unit length of the string (measured in $kg\,m^{-1}$). The different frequencies of stationary wave are called harmonics, with the lowest frequency harmonic called the first harmonic. Stationary waves can also be demonstrated easily in the laboratory using sound and microwaves. These are usually produced when the waves reflect off a barrier and the reflection and incident waves interfere with each other, forming the stationary wave.

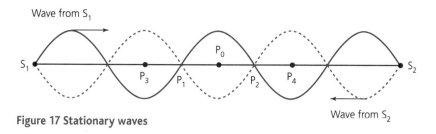

Figure 17 Stationary waves

Required practical 1

Investigation into the variation of the frequency of stationary waves on a string with length, tension and mass per unit length of the string

The diagram below shows a simple way of investigating stationary waves on a string.

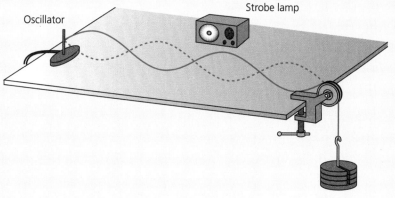

Figure 18 Investigating stationary waves on a string

With this experimental apparatus it is possible to measure the frequency of stationary waves on the string by altering the frequency of the oscillator and using the strobe lamp to 'freeze' the motion of the string. It is then possible to vary the tension and the length of the string easily, and the mass per unit length can be altered by changing the string.

5 An oscillator produces a stationary wave of frequency *f* on a stretched string, under a tension *T*. The tension on the string is then quadrupled to 4*T*. Which change listed below will produce the same frequency of stationary wave, *f*? (AO2)

1 mark

   A   doubling the length

   B   halving the length

   C   doubling the mass per unit length

   D   halving the mass per unit length

6 A guitar string of mass 2.3 g and length 80 cm is fixed onto a guitar. The string is tightened to a tension of 78 N between bridge A and bridge B, a distance of 60 cm, as shown in the diagram.

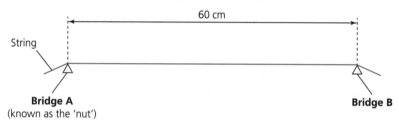

Figure 19 A guitar string

Calculate the frequency of the stationary wave harmonic produced when plucking this string. (AO2)

Exam-style questions

(27)

1 Earthquakes produce both transverse and longitudinal seismic waves that travel through the rock. Figure 20 shows the vertical displacement of the particles of a rock at a given time, for different positions along a transverse wave.

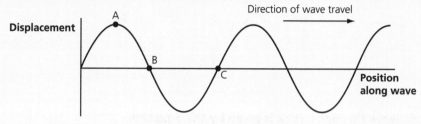

Figure 20 A seismic wave

a State the phase difference between: **3 marks**

i points A and B on the wave

....

ii points A and C on the wave

....

iii points B and C on the wave

....

b Describe the movement of a rock particle at point B as the next complete wave cycle passes through it. **2 marks**

....

....

....

....

c Some seismic waves arrive at a detector having undergone polarisation. State and explain what can be deduced about the wave from this information. **2 marks**

....

....

....

d The frequency of the seismic wave is measured to be 5.5 Hz.

i State what is meant by the frequency of a progressive wave. **1 mark**

....

....

ii Calculate the wavelength of the wave if its speed is $4.8 \times 10^3 \, \text{m s}^{-1}$, giving your answer to an appropriate number of significant figures. **2 marks**

....

....

....

....

e Transverse seismic waves are called secondary or s-seismic waves because they are slower than the faster primary or p-seismic waves, which are longitudinal waves and arrive first at a detector after an earthquake.

i Describe the motion of a rock particle at point B when the p-wave travels through it. **2 marks**

....
....
....
....

f The speed of a p-seismic wave through a rock is given by the equation:

$$v_p = \sqrt{\frac{M}{\rho}}$$

where M is the rock elasticity and ρ is the density of rock. Calculate the speed of a p-wave through a granite rock with $M = 67.5\,\text{GPa}$ and $\rho = 2.7\,\text{g cm}^{-3}$. **3 marks**

② The diagram below shows a stationary wave on a fixed string. The string is tied onto a thin metal bar at A and clamped at B. An oscillator causes the bar to oscillate up and down at a given frequency:

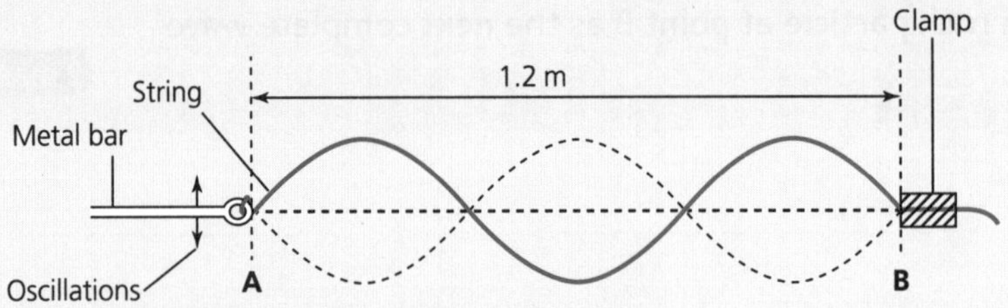

Figure 21 Stationary waves in a vibrating string

Explain how a stationary wave is formed on this string. Then describe the key features of the stationary wave shown in the diagram above.

The quality of your written answer will be assessed in this question. **6 marks**

....
....
....
....
....
....
....
....
....
....

Topic 2 Refraction, diffraction and interference

Interference

Path difference and wave superposition

Two wave sources are **coherent** if they have the same frequency and emit waves in phase or with a constant phase difference. A source is **monochromatic** if it emits waves with the same (constant) frequency and wavelength. Two such sources, S_1 and S_2, emit coherent waves towards a detector at O, as shown in the diagram below:

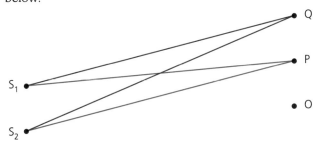

Figure 22 Path difference and the superposition of waves

The distance S_1O is the same as distance S_2O, and so the waves from each source have the same path distance and the **path difference** ($S_1O - S_2O$) is zero. The waves arrive in-phase, crest meets crest, trough meets trough and the waves constructively interfere. The waves arriving at P have a different path difference ($S_1P - S_2P$). If this path difference is equal to an odd number of half-wavelengths, then crest meets trough and the waves will **destructively interfere** and a detector at P will detect a minimum amplitude wave. **Constructive interference** occurs at every point such as Q, where the path difference is equal to an integral number of whole wavelengths. Superposition of waves can be shown in the laboratory using the sound from two speakers in parallel or by light passing through a double slit.

Interference and diffraction of light

Interference and diffraction can be easily demonstrated in the laboratory using a laser as the source of monochromatic light. A modern version of an experiment first conducted by Thomas Young in 1803 involves light from a laser passing through a double slit producing an interference pattern as a series of bright and dark fringes on a screen. The fringe spacing, w, can be measured on the screen, and is given by the Young's double-slit equation:

$$w = \frac{\lambda D}{s}$$

where λ is the wavelength of the monochromatic laser light, D is the slit–screen distance and s is the slit separation. A polychromatic white light source, such as the Sun, or a light bulb, produces a similar effect, but the interference fringes consist of spectra of different colours.

Thomas Young's experiment showed that light behaves as a wave in this experiment, confirming a view first expressed by Christiaan Huygens in 1678, which was then contradicted by Isaac Newton in 1704, who proposed a particle model of light. Light was finally characterised as having wave–particle duality in the early part of the twentieth century.

Lasers have safety risks associated with their use. Although the power of lasers used in schools and colleges is usually quite low, the light is monochromatic, coherent and confined to a small beam area, so if it hits the retina of an eye it can cause some damage. When a laser is being used, laser goggles should be worn and a laser hazard symbol should be displayed on the entrance to the laboratory.

Required practical 2

Investigation of interference effects to include the Young's slit experiment and interference by a diffraction grating

The diagram below shows a simple way of investigating the interference of light using a laser and a double slit:

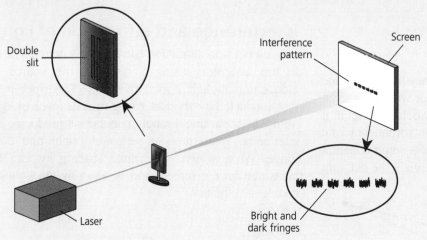

Figure 23 Investigating the interference of light using a laser and a double slit

With this experimental apparatus it is possible to investigate the Young double-slit equation and the variation of the fringe spacing (*w*) on the screen. The slit–screen distance and the slit separation can be varied, and their effect on the fringe spacing observed and measured.

---

1 Which of the following changes will *not* double the fringe spacing of a Young's double-slit diffraction pattern? (AO1)　　`1 mark`

   A　doubling the wavelength

   B　doubling the slit–screen distance

   C　doubling the slit separation

   D　halving the slit separation

2 A laser beam is described as coherent and monochromatic.

   a　Explain what is meant by the terms:

     i　coherent (AO1)　　`1 mark`

     ii　monochromatic (AO1)　　`1 mark`

b **Describe and explain one safety precaution that should be taken when using a laser to perform diffraction experiments. (AO1)** (2 marks)

....

....

....

....

c **A laser produces light with a wavelength of 632.8 nm. The laser beam strikes a pair of slits, 0.20 mm apart. A diffraction pattern is produced on a screen 4.6 m away from the slits. Calculate the fringe spacing on the screen. (AO2)** (3 marks)

Diffraction

Single-slit diffraction

When laser light passes through a single narrow slit, the light diffracts off each edge of the slit and produces a characteristic diffraction pattern on a screen as shown below:

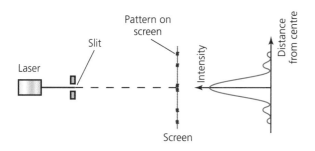

Figure 24 Single-slit diffraction

A similar pattern occurs with polychromatic white light, where a spectrum of colours is produced. As the wavelength of the light approaches the same size as the slit width, so the width of the central maximum is greatest and the diffraction pattern is most spread out. The diffraction pattern narrows as the difference between the wavelength and the slit width increases.

Diffraction gratings

Diffraction gratings consist of many narrow slits close together. When laser light is passed through a diffraction grating, each slit acts as a coherent source and a diffraction pattern is produced on a screen. The pattern consists of a series of spots called 'orders' on the screen, centred away from a central bright zero order. The orders are symmetrical either side of the central zero order and number away from zero. The whole diffraction pattern is described by the diffraction equation:

$$n\lambda = d \sin \theta$$

where n is the order number, λ is the wavelength of the light, d is the slit spacing and θ is the angle between the zero order and the nth diffraction order. Diffraction gratings are used in spectrometers to analyse the light coming from light-emitting objects such as stars and furnaces. CDs and DVDs show a diffraction effect when light reflects off them. The structure of crystals can be determined by using X-rays, where the atomic structure of the crystals act as diffraction gratings.

3 A single-slit diffraction pattern is produced by a laser beam on a screen. The laser has a wavelength of 600 nm and the single slit has a width of 0.9 µm. The slit width is decreased to 0.7 µm. Which of the following changes are observed on the screen? (AO1)

1 mark

	Effect on the central diffraction maximum	Effect on the brightness of the pattern
A	Widens	Darkens
B	Narrows	Brightens
C	Narrows	Darkens
D	Widens	Brightens

4 A diffraction grating with 330 lines per mm produces a series of diffraction orders on a screen 5.4 m from the grating. The wavelength of the laser light is 632.8 nm. Calculate the distance between the first and second diffraction orders as measured on the screen. (AO2)

4 marks

Refraction at a plane surface

Refraction

Refraction occurs when waves travel from one medium, where they travel at one speed, into another medium where they travel at a different speed. The change of speed causes a change in wavelength, and, if the waves are travelling at an angle to the boundary between the two media, then the wave will change direction. The property of the substance that dictates the change of speed is called the **refractive index (n)** of the substance, given by the equation:

$$n = \frac{c}{c_s}$$

where c is the speed of light in a vacuum, and c_s is the speed of the light in the substance. For air, where the speed of light is close to the speed of light in a vacuum, the refractive index is approximately 1.

The relationship between the angle of incidence and the angle of refraction for a refracting wave is given by Snell's law:

$$n_1 \sin \theta_1 = n_2 \sin \theta_2$$

where n_1 and θ_1 are the refractive index and the angle of incidence in medium 1; n_2 and θ_2 are the refractive index and the angle of refraction in medium 2.

Total internal reflection

When light travels from a medium with a high refractive index (e.g. glass) into a medium with a low refractive index (e.g. air), the beam refracts away from the normal line, as shown in Figure 25a.

With increasing θ_1, eventually an angle of incidence is achieved where the refracting beam refracts at an angle of 90° to the normal, along the boundary between the two media, as shown in b). The angle of incidence at this point is called the **critical angle** of the medium (θ_c). For angles of incidence greater than the critical angle, **total internal reflection** occurs and the beam is reflected back into the medium, as shown in c). The critical angle between two media is given by:

$$\sin \theta_c = \frac{n_2}{n_1}$$

where n_1 and n_2 are the refractive indices of medium 1 and medium 2.

Fibre optics

Total internal reflection is used in fibre optics. Signals of light or infrared are passed down thin strands (or fibres) of transparent materials such as glass and plastics. The beams totally internally reflect off the inside surfaces of the fibres, which are coated with a thin outer cladding. This has a lower refractive index than the material of the fibre, which ensures total internal reflection, provided that the incident beam is introduced into the fibre at an angle greater than its critical angle.

Optical fibres suffer from dispersion of the signal. Different wavelengths (colours) of light (or infrared) travelling through glass slow down by different amounts because the refractive index varies with wavelength, and the different wavelengths travel along slightly different paths. As a signal travels within an optical fibre, it therefore disperses and spreads out. Two types of dispersion occur in step-index optical fibres:

- **material dispersion** occurs because the refractive index of the optical fibre varies with wavelength. If the signal consists of a sharp pulse of light containing slightly different wavelengths, the different wavelengths travel at slightly different speeds, causing the sharp pulse to spread into a broader signal. The duration of each pulse increases **(pulse broadening)** and this limits the maximum frequency of pulses and therefore the bandwidth available for use in the fibre. Using monochromatic light (or infrared) reduces this effect.

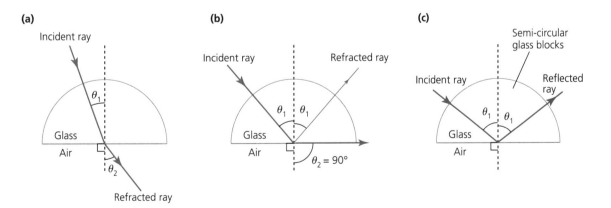

Figure 25 Total internal reflection of light

- **modal dispersion** occurs when rays inside the optical fibre take slightly different paths. Rays taking longer paths take longer to travel through the fibre, so the duration of the pulse increases and the pulse broadens again. Narrow fibres reduce this effect.

Some wavelengths of light (or infrared) are absorbed strongly by the materials used to make optical fibres, so the signal strength reduces. Optical fibres are manufactured from materials with low absorption at the wavelength used to send optical signals.

5 The material of an optical fibre is chosen to match the wavelength of the signal travelling down it. Which of the following changes will not reduce the dispersion of a signal in the optical fibre? (AO1) `1 mark`

   A Reduce the time period of the pulse.

   B Use a monochromatic signal.

   C Make the fibre narrower.

   D Change the refractive index of the fibre.

6 A Perspex (plastic) cube floats on top of a pool of oil and a beam of light is incident onto the vertical face of the cube. The angle of incidence is 41° as shown in Figure 26. The angle of refraction in the Perspex is 26° and the ray is totally internally reflected at P.

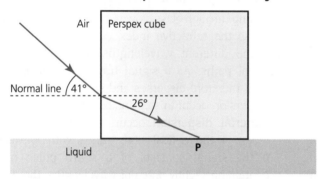

Figure 26 Refraction by a Perspex cube

   a Complete the path of the beam through the cube to the air again. (AO3) `3 marks`

   b Show that the refractive index of Perspex is 1.49. (AO2) `2 marks`

Exam-style questions

1 a The basic wave equation relates wave speed, frequency and wavelength and is given by:

$$c = f \times \lambda$$

Describe how each of these quantities change, if at all, when light travels from a vacuum into clear Perspex.

3 marks

c ....

f ....

λ ....

The diagram below shows a cross-section through a step-index optical fibre.

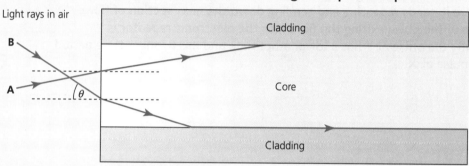

Figure 27 An optical fibre cross-section

b Beam A is incident at the end of the optical fibre, travels through the air–core boundary and then experiences total internal reflection. On the diagram, show the path of this ray down the fibre.

2 marks

c i The speed of light in the core material is $2.01 \times 10^8\,\mathrm{m\,s^{-1}}$. Calculate the refractive index of the core material.

2 marks

ii The refractive index of the cladding is 1.44. Calculate the critical angle of the core-cladding boundary.

2 marks

d Beam B is incident on the end of the fibre, refracts through the air–core boundary and then refracts again at the core-cladding boundary, travelling along the boundary. Calculate the angle of incidence, θ, of this beam at the air–core boundary.

3 marks

e A monochromatic pulsed light signal (X) can travel down 10 km of optical fibre before it needs repeating and sending down the next section of fibre. The shape of the pulse exiting the fibre into the electronic repeater is Y. Explain why the pulse at Y has a lower amplitude and has a longer time period than the pulse at X.

2 marks

....

....

....

....

2 a A diffraction grating arrangement is used to carry out measurements in order to calculate the wavelength of light from a college laboratory laser. A grating of diffraction lines of a known spacing (number per mm) is used to produce a diffraction pattern on a screen. Draw a fully labelled diagram of the experimental arrangement, describe the measurements that need to be taken and then explain how the measurements can be used to calculate the wavelength of the light. State two safety precautions that could be taken to reduce the risks associated with this experiment and state how you would set up the equipment to improve accuracy.

The quality of your written communication will be assessed in this question.

6 marks